THE ELECTROMAGNETIC SPECTRUM



# UNDERSTANDING AND USING RADIO WAVES

ELIZABETH RUBIO

Enslow PUBLISHING

Published in 2022 by Enslow Publishing, LLC.
101 W. 23rd Street, Suite 240, New York, NY 10011



**Library of Congress Cataloging-in-Publication Data**

Names: Rubio, Elizabeth, author.
Title: Understanding and using radio waves / Elizabeth Rubio.
Description: New York, NY : Enslow Publishing, LLC., [2022] | Series: The electromagnetic spectrum | Includes bibliographical references and index. | Audience: 7-12.
Identifiers: LCCN 2019013435| ISBN 9781978514867 (library bound) | ISBN 9781978514850 (pbk.)
Subjects: LCSH: Radio waves—Juvenile literature.
Classification: LCC QC676.4 .R83 2022 | DDC 537.5/34—dc23
LC record available at https://lccn.loc.gov/2019013435

Printed in the United States of America

To Our Readers: We have done our best to make sure all websites in this book were active and appropriate when we went to press. However, the author and the publisher have no control over and assume no liability for the material available on those websites or on any websites they may link to. Any comments or suggestions can be sent by email to customerservice@enslow.com.

Photos Credits: Cover Titima Ongkantong/Shutterstock.com; p. 5 FilippoBacci/E+ /Getty Images; pp. 8, 9, 18 VectorMine/Shutterstock.com; p. 11 Jeff Greenberg /Universal Images Group/Getty Images; p. 15 Sergey Merkulov/Shutterstock.com; p. 21 lilly_fai/Shutterstock.com; p. 23 Designua/Shutterstock.com; p. 27 PhotoQuest/Archive Photos/Getty Images; p. 31 Fouad A. Saad/Shutterstock.com; p. 32 Viktorus/Shutterstock.com; p. 34 X-ray: NASA/CXC/UCLA/Z. Li et al; Radio: NRAO/VLA; p. 37 FooTToo/Shutterstock.com; p. 39 ESO; p. 41 Space Frontiers/Archive Photos/Getty Images; p. 43 Big Ear Radio Observatory and North American AstroPhysical Observatory (NAAPO)/Wikimedia Commons/File: Wow signal.jpg/Public Domain; p. 48 Natalia Golubnycha/Shutterstock.com; p. 53 The vyhuhol/Shutterstock.com; cover background, interior pages (faceted color pattern) s_maria/Shutterstock.com; interior pages (wave graphic) MagicMary /Shutterstock.com.

# CONTENTS

# INTRODUCTION

What comes to mind when hearing the word "radio"? Maybe a favorite song playing in the car? Perhaps the channel you most often tune to? Maybe even a set of walkie-talkies?

What most people think of as a "radio" is actually a radio receiver. It is a device that absorbs radio waves and translates their signal into sound. So just what are radio waves?

Radio waves are actually a form of light. This doesn't mean you can see them. In fact, radio waves have much lower energy than the light we can see. But, just like visible light, radio waves are given off by energetic objects. These objects can be radio broadcasting antennas, but they can be other things, too.

Humans have harnessed the power of this invisible light to communicate across the globe. Radio waves travel at the speed of light, meaning radio transmissions pass from the transmitter to the receiver so fast humans perceive it as instantaneous. The speed of light is nearly 186,282 miles (300,000 kilometers) per second, which means a single radio wave can cross the entire Earth in less than two-tenths of a second. This is why it is possible to have a conversation with someone via a two-way radio and not have to wait between sentences.

People use radio waves to broadcast not just sound but also images. That's right—even television signals are sent via radio waves! An antenna that receives a television broadcast must be connected to a device that can interpret the signal and display the picture in addition to the sound.

Radio communications require both a transmitter and a receiver. Radio transmission towers like this one broadcast signals that can be picked up by receivers miles away.

Radios that are used to talk to each other, such as walkie-talkies and CB radio, also rely on radio waves. People also use radio waves to access the internet. The Wi-Fi signal in a home or school is made of radio waves. The signal a mobile phone uses to make calls and transmit data is a radio wave. Radio waves are vital to many forms of communication.

Another surprising use of radio waves is the study of the galaxy. Stars emit all kinds of light: visible light, ultraviolet light, and, yes, radio waves. Scientists can receive these radio waves and interpret them to learn more about galactic objects, such as stars, black holes, and gas clouds.

It has been less than two hundred years since radio waves were discovered, but humans rely on them for many tasks. For the most part, these tasks involve communication. As science advances, even more new uses for radio waves may be discovered.

# THE ELECTROMAGNETIC SPECTRUM

CHAPTER ONE

To understand radio waves, it is important to understand the electromagnetic spectrum. All light is part of the electromagnetic spectrum. Another term for light is "electromagnetic radiation." "Radiation" may sound scary, but it's actually why radio waves got their name—they radiate away from their source. Not all electromagnetic radiation is harmful. Radio waves aren't harmful. In fact, radio waves are all around you!

The electromagnetic spectrum includes the entire range of electromagnetic radiation. This radiation includes all forms of light—not only radio waves, but also visible light, microwaves, X-rays, and gamma rays. Light is also called electromagnetic radiation because it is a wave of alternating electric and magnetic fields. Just like when you drop a stone into water, light travels in waves, spreading out in all directions from its source.

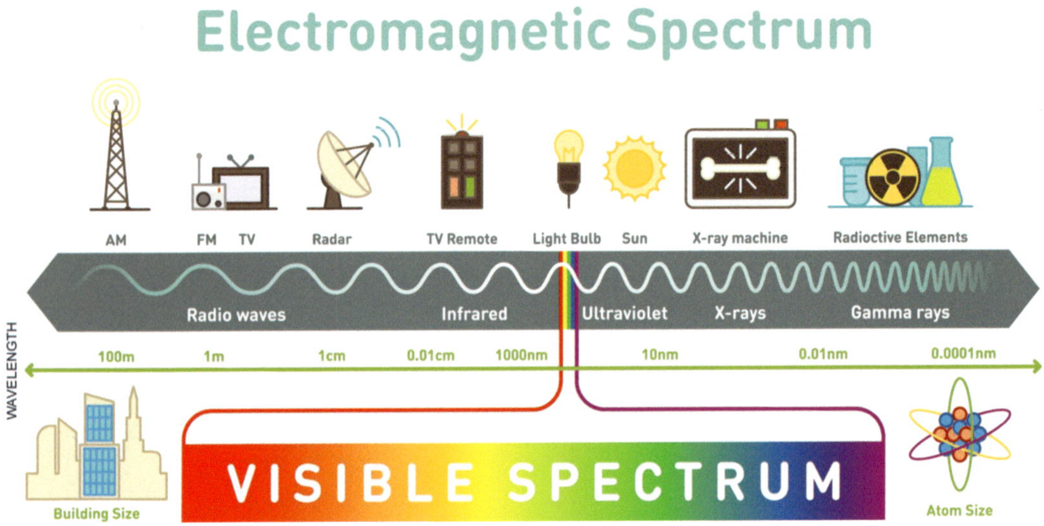


The electromagnetic spectrum encompasses all kinds of light, from gamma rays to radio waves. Radio waves lie at the end of the spectrum; they are the light with the lowest energy.

All types of light are waves, so each type of light has a specific wavelength. A wavelength is the distance between two crests of a wave. Light with shorter wavelengths has more energy than light with longer wavelengths. Put another way, light with shorter wavelengths is more powerful and can penetrate more objects than light with longer wavelengths.

Wavelengths in the electromagnetic spectrum range from thousands of miles (kilometers) on the low-energy end of the spectrum to $1 \times 10^{-13}$ meters on the high-energy end. Put another way, the lowest energy light has a wavelength longer than a country, and the highest energy light has a wavelength smaller than the diameter of an atom.

# How Do Radio Waves Compare to Other Types of Electromagnetic Radiation?

Radio waves lie on the end of the electromagnetic spectrum with very long wavelengths. They encompass a huge range of wavelengths, from 0.04 inch (1 millimeter) to 6,200 miles (10,000 km).[1] That means the smallest radio wavelengths are about the size of the head of a pin, while the longest ones are about the distance between Los Angeles, California, and Rome, Italy!

Most radio waves humans use are on the shorter end of that spectrum, between 0.04 inch (1 mm) and 62 miles (100 km).[2]

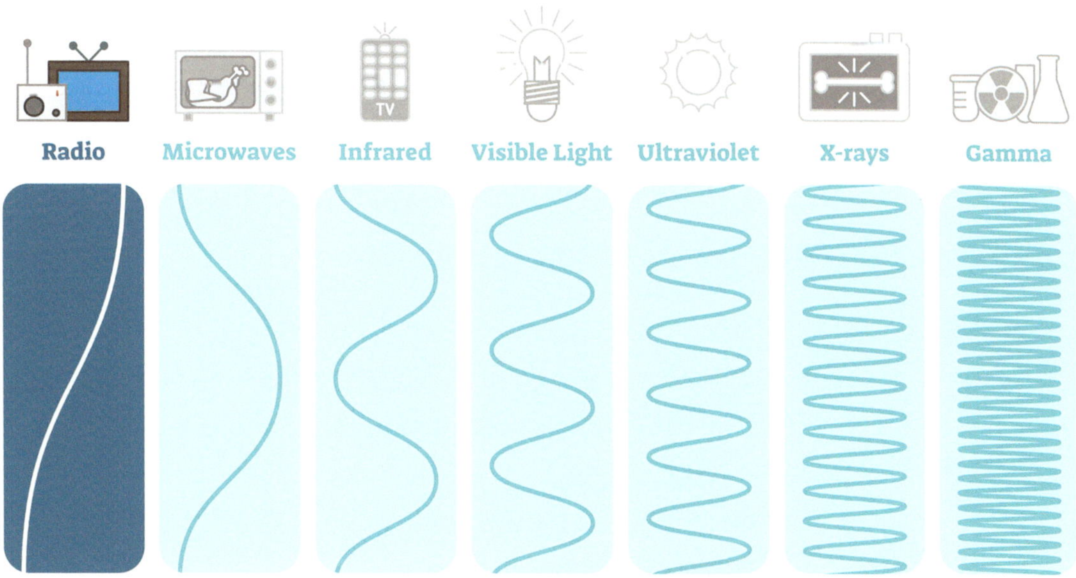


Radio waves have the longest wavelengths of any type of electromagnetic radiation. Wavelength and energy are inversely related, which means that long wavelengths have low energy.

The distance between Los Angeles and Ventura, both in California, is about 62 miles (100 km). Radio waves with wavelengths longer than 62 miles (100 km) have very limited use because the human voice cannot be encoded in these waves. However, there is some use for these waves, as they can travel through salt water, while radio waves with shorter wavelengths cannot.

Radio waves are non-ionizing radiation, meaning they cannot eject an electron from an atom. This makes them harmless to living things. Ionizing radiation, which includes electromagnetic radiation of wavelengths shorter than those in the visible spectrum, can eject electrons from atoms and can therefore be harmful. These waves include X-rays, ultraviolet light, and gamma radiation. This harmful potential of ionizing radiation is why you wear sunscreen when you are out in sunlight, which includes ultraviolet light. You need protection from these types of rays so you are not harmed. But radio waves can and do exist all around you without harming you in the least. No amount of exposure to radio waves is dangerous.

## Emission of Radio Waves

Radio waves, like visible light, are a form of light. Light is emitted after electrons in an atom become excited. When these electrons return to their normal, unexcited state, light is given off. Additionally, heat is given off, which is why all objects that emit light also emit heat. The amount of heat given off varies. Because radio waves have very little energy, the amount of heat associated with them is quite small. A person cannot feel the heat associated with radio waves because the human body gives off far more heat than objects that emit radio waves.

# TRACKING STORMS WITH RADIO WAVES

**The National Oceanic and Atmospheric Administration (NOAA) uses Doppler radar to track storms. This type of radar sends a radio signal from a single point and measures the way it reflects off of moving objects. Objects moving toward the source of the radio waves will increase its frequency, while those moving away from the source will decrease its frequency. In the case of storms, the objects NOAA measures are raindrops.**

NOAA storm tracking relies on radar, which stands for RAdio Detection And Ranging. Any time you see a radar image of a storm on a weather app, remember it was made using radio waves!

Very few things on Earth give off radio waves naturally. One of these things is lightning. When lightning strikes, a burst of radio waves is emitted. You may have noticed that your cell phone or television is subject to interference during a lightning storm, which might either be heard as crackles or might be so severe your signal is lost. This effect is because the radio waves from the lightning are interfering with the radio waves that allow your cell phone to communicate with a cell tower or that allow your television to communicate with the broadcasting station.

In fact, the NOAA tracks the progress of lightning storms using radio receivers. These receivers are spaced at known distances from each other. They absorb the radio waves generated by lightning at slightly different times depending on how far they are from the lightning strike. By measuring these differences, NOAA is able to pinpoint the location of a lightning strike. It can map these strikes over the course of several minutes to determine the direction and speed of a moving lightning storm. This helps prepare people in the storm's path for dangerous weather.[3]

Although very little natural emission of radio waves occurs on Earth, a great deal of artificial emission occurs, now that humans have discovered how to send and receive radio signals. Take a look on your next car trip and you might notice tall towers topped with flashing lights and, in some cases, curved dishes. These are radio broadcasting towers and cellular towers, respectively. These towers help people communicate using radio waves.

CHAPTER TWO

# THE DISCOVERY OF RADIO WAVES

Radio waves were predicted before anyone actually confirmed they existed. The prediction of radio waves arose as scientists began to understand that light, electricity, and magnetism were all different aspects of the same phenomenon. Scientists knew about visible light, and they suspected there was more to the picture than they could see.

## Maxwell's Prediction

James Clerk Maxwell was a Scottish physicist who lived from 1831 to 1879.[1] Maxwell linked electricity and magnetism and coined the term "electromagnetic radiation." He derived equations that described electric and magnetic fields. He used his equations to build a model that predicted the speed of waves of electromagnetic energy and discovered that the speed he had predicted was nearly identical to the speed of light.[2] This confirmed his idea that electricity, magnetism, and light were all different aspects of the same phenomenon.

In 1873, Maxwell published a book called *Treatise on Electricity and Magnetism*. This book included the equations Maxwell had derived and data generated by his model. In the book, Maxwell concluded that light consisted of transverse "undulations," or waves, and was made of the same stuff as electricity and magnetism.[3] Later experiments would show this to be true.

Maxwell thus predicted that electromagnetic waves existed, although he could not detect them experimentally at the time. This was because instruments to detect weak electromagnetic fields had not yet been developed. Furthermore, Maxwell's equations predicted wavelengths below infrared, which were the lowest wavelengths that had, at the time, been observed.[4] Maxwell's equations indicated that light at much larger wavelengths existed, as well as very short wavelengths. This was only mathematical theory at the time because instruments were still unable to detect very short or very long wavelengths.

## Hertz's Discovery

German physicist Heinrich Hertz, who lived from 1857 to 1894, set out to experimentally show that Maxwell's ideas about light were correct. He reasoned that if electromagnetism existed as waves, it could travel between devices.

Hertz built an experimental setup to demonstrate the existence of electromagnetic waves. His setup consisted of a coil of wire connected to a battery. The battery was connected to a switch. At the other end of the apparatus was a pair of rods with a tiny gap between them. These rods were connected to the battery. Positioned several yards from this apparatus was a receiver loop, a loop of wire with a small gap, similar to the gap in the apparatus. This loop was not connected to any kind of power source.

Hertz considered that, if electromagnetism traveled in waves as predicted by Maxwell's equations, when the circuit on the apparatus was closed, not only would a spark jump across the gap on the apparatus, but another spark would jump across the gap on the receiver loop. This would happen because the

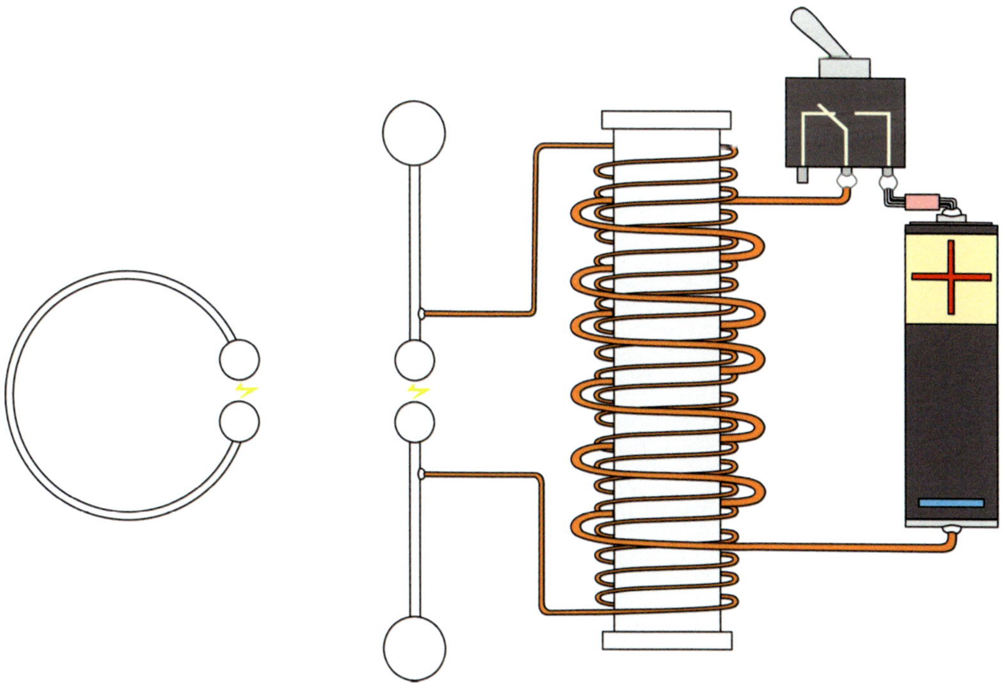

The apparatus Hertz used to demonstrate the existence of electromagnetic waves consists of very simple components. Many people have built and tested their own versions.

receiver loop was in the path of the waves of electromagnetism generated by the apparatus.

Sure enough, when Hertz fired up the apparatus, sparks jumped across the gaps in both the apparatus and the receiving loop. They were not bright sparks. In fact, Hertz had to perform his experiment in a dark room in order to see the sparks at all. Yet they definitely did exist. For the first time in history, radio waves had been experimentally observed.[5]

In 1888, Hertz published his findings. Before long, his groundbreaking research would be used by people all over the world to build radio transmitters and receivers, revolutionizing the communications of modern humans.

Hertz's setup, though revolutionary at the time, is so simple in its design that it can be replicated by anyone with a basic understanding of electricity and a few simple materials. Many scientists have built their own versions, verifying Hertz's conclusions time and time again.

## The Hertz Unit

Light waves can be measured in terms of wavelength. They can also be measured in terms of frequency, or how many waves pass through a given point in a given time. Because light travels at the same speed no matter its wavelength, frequency is commonly used to describe light waves, especially radio waves.

Because of Hertz's groundbreaking experiments, the unit of frequency of light waves is named for him. One hertz (Hz) is equal to one wave per second. Because light frequencies range from very small to very large, prefixes are sometimes added to

## SOUND AND HERTZ

**Sound waves, like radio waves, can be measured in hertz. For sound waves, frequency is associated with pitch. Low frequencies are associated with sounds that have low pitch. High frequencies are associated with sounds that have high pitch.**

**The lowest frequency humans can hear is about 16 Hz in normal air. The highest frequency is about 20 kHz in normal air. As humans age, the ability to hear sounds at either end of this range decreases. The human speaking voice is usually between 500 and 1,000 Hz.**

the unit Hz. These prefixes are the standard prefixes used in the metric system. For example, one kilohertz (kHz) is equal to 1,000 Hz. One megahertz (MHz) is equal to 1,000,000 Hz.

For light, a frequency of 1 Hz is equal to a wavelength of 186,282 miles (299,792 km). Importantly, Hz can be used to measure oscillations other than light waves. Most commonly, the unit Hz is applied to light waves (especially radio waves) and sound waves. The speed of a sound wave depends on the material through which it moves, so the measurement of a sound wave in hertz is not easily converted to a wavelength unit.

Wavelength and frequency are inversely related. This relationship means the larger one is, the smaller the other is. Radio waves with very high frequencies have very small wavelengths, while radio waves with very low frequencies have very large wavelengths. This is due to the different aspects of the wave that each unit measures. The wavelength is the distance

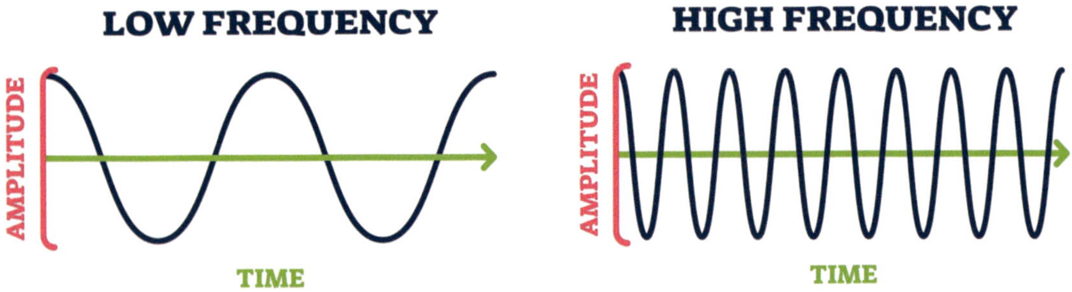


As wavelength decreases, frequency increases. Frequency is the number of waves that pass a point in a given time. Amplitude refers to the wave's height.

between two peaks of a wave. Because light travels at a constant speed, waves with shorter wavelengths pass through a point more times in a second than waves with longer wavelengths.

To put it another way, imagine a bus and small cars. Three of the cars together are as long as the bus. If they all move at the same speed, three cars will pass a single point in the same amount of time it takes the bus to pass that point. If you think of the length of each vehicle as a wavelength and the number of times the vehicle can pass a single point in a given amount of time, the relationship between wavelength and frequency begins to make sense.

# TRANSMITTING AND RECEIVING RADIO WAVES

CHAPTER THREE

Much of the use of radio waves today involves communication. Radio waves are used to broadcast music, speech, and information through the air. How is this accomplished?

## Radio Devices

Radio communication requires devices. Namely, a transmitter must send out radio waves, and a receiver must absorb them and translate the information encoded in them.

Radio transmitters require a power source, a modulator, and an antenna. The power source generates the electricity needed to operate the transmitter. The modulator encodes the information in the radio waves. The antenna sends the radio waves out into the space around the transmitter.

Radio receivers require an antenna, a demodulator, and generally an amplifier. Depending on the type of receiver used, they may also require a power source. The antenna absorbs the radio waves from the space around it, exactly opposite the function of the antenna in the transmitter. The demodulator decodes the information encoded by the modulator in the transmitter. Finally, the amplifier increases the signal. Once, vacuum tubes were commonly used as amplifiers, but modern radios usually rely on transistors instead. Both of these amplifiers increase the electrical signal in the radio waves. Most receivers also require a power source, such as a plug or batteries. Many receivers also contain a tuner, which can be adjusted to pick up only a single specific radio wave frequency.

Radio antennas often appear as long, thin metal rods. For example, most cars have a radio antenna somewhere on the outside of the vehicle. Some antennas are curved dishes with a small pin facing the dish. The curvature of the dish helps capture and focus radio waves, especially weak ones. The pin collects these focused waves and sends the signal along a wire to a receiver. You may have noticed these dishes on homes attached to an exterior wall or roof. These dishes receive radio signals from satellites orbiting Earth. Look closely the next time you see one of these dishes and you will notice the wire running from the dish into the house. This is the wire that connects the antenna to the receiver.

Aside from the AM/FM radios most people are familiar with, there are many other radio types. Police radios are limited channel radios through which police and emergency personnel communicate with each other. Similarly, military radios are limited channel radios for use by the military. CB radios, or citizens band radios, are short-distance transmission radios

The antenna on this car allows the car's radio to receive AM/FM broadcasts. What other types of antennas have you seen?

used only for noncommercial purposes. They are also called ham radios. Short-range radios in which the devices are both transmitters and receivers (called transceivers) are called walkie-talkies, and they can be used by anyone for communicating across short distances. Most long-range radio usage is regulated by government agencies and is subject to licensing, but most short-range radio usage can be done without a license.

## HAM RADIO

**Ham radio is the use of radio transmissions by unlicensed civilians. Ham radio has its own special terms that operators use to communicate quickly and effectively. Most shipping vehicles, such as lorries and tractor trailers, use ham radio to talk to each other about their routes and deliveries. Many people use ham radios in their homes.**

**Anyone can buy and use a ham radio—even you! Ham radio operators across the globe talk to each other every day. Even if you don't know the special terms, the ham radio community will welcome you and help you learn.**

## Transmission, Modulation, and Reception

Radio transmitters send radio waves into the air. Radio receivers absorb them and relay the information. But just how is information encoded in radio signals, and how does it travel?

Modulation is the process in which information is encoded in radio signals. Modulation describes changing an aspect of the wave. There are two main types of radio modulation: amplitude modulation (AM) and frequency modulation (FM).[1]

Amplitude refers to the size of a wave. Frequency refers to how many wave oscillations occur in a second. In amplitude modulation, the height (amplitude) of the wave is changed. In frequency modulation, the number of waves that pass a given point in a second is changed.

# WAVE SHAPES

Amplitude modulation (AM)

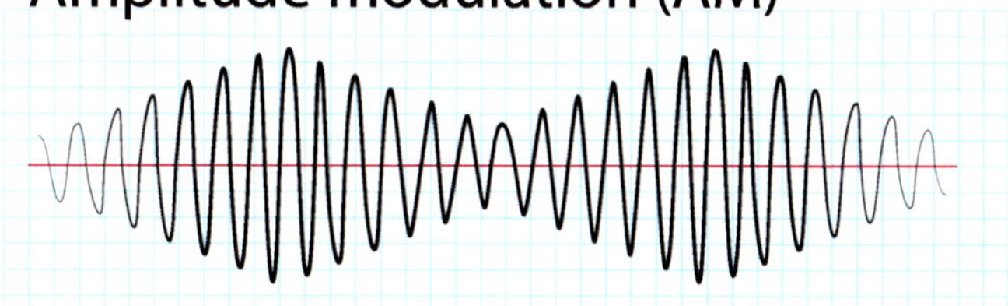

Frequency modulation (FM)

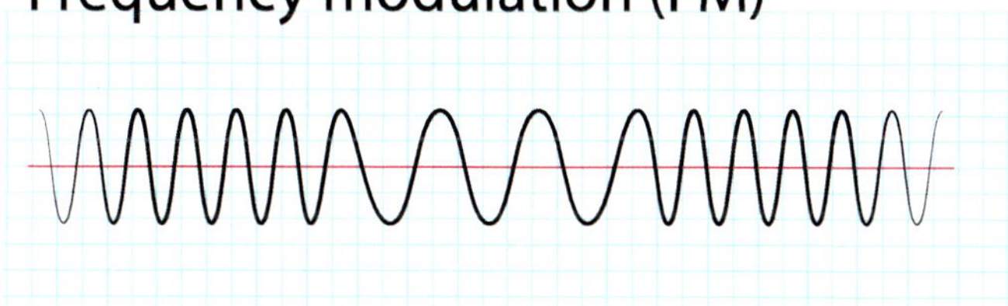

Carrier signal

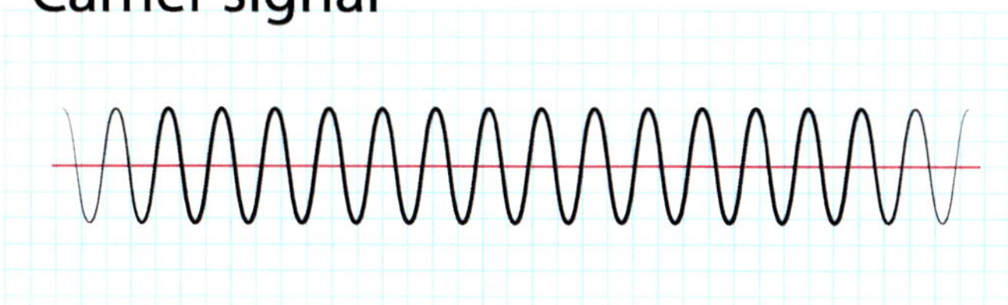

In AM, the height of the wave is changed in order to encode information. In FM, the frequency of the wave is changed. The radio signal that is changed is called the carrier signal.

The information encoded in a radio signal exists within the modulation itself. The radio signal that travels from the transmitter to the receiver simply exists to carry the information it encodes. For this reason it is called a carrier. The specific frequency of the carrier allows the receiver to filter out other signals so a single signal can be detected. But any information can be sent on any frequency, within a few limits. For example, the human voice cannot be sent on very low frequencies.

An important aspect of radio waves is that their very long wavelengths often allow them to pass through objects that reflect or absorb other types of light. This is why radio receivers can pick up signals inside buildings and vehicles. However, radio signals can be absorbed by some materials, and a large enough amount of any material is likely to absorb them eventually. If your radio signal has ever faded out when you drove through a tunnel, you will be familiar with this phenomenon. This is why many vehicles and buildings have external antennas. These antennas can pick up the signal before it is blocked by materials such as wood or metal and transmit the signal along wires to the receiver.

# RADIO WAVES IN COMMUNICATION

CHAPTER FOUR

Radio waves are most commonly used in communication. Even before radio waves were actually shown to exist, people hoped to find ways to communicate across distances. Since 1901, people have used radio waves to accomplish this communication.[1]

## The History of Radio

In the late nineteenth century, scientists and inventors made new discoveries and inventions all the time. Communication was a big focus for many of these people, and the telephone, patented by Alexander Graham Bell in 1876, had finally made this possible.[2] But many people, including Bell, had noticed that sometimes their telephones picked up sparks and interpreted them as noise. Most of these people dismissed this as an anomaly, or possibly some sort of electrical problem. Italian inventor Guglielmo Marconi thought there was more to it. Building on the ideas put forth by Heinrich Hertz and several others, he built a radio transmitter and receiver. With his earliest designs, he was able to send and receive messages at a distance of about a mile and a half. While this accomplishment was groundbreaking, it was not enough for Marconi. He refined and improved his designs. In 1901, he received a message in Newfoundland that had been sent from England—2,200 miles (3, 540 km) away![3] In 1906, Lee de Forest invented a vacuum tube that amplified radio signals, which was the first use of amplitude modulation. That same year, Reginald Fessenden broadcast a radio transmission featuring speech and music. Radio broadcasting was born. In 1912, the United States began regulating and licensing use of radio wave frequencies.[4]

By the 1920s, commercial radio broadcasting was on the rise. Many newspapers were among the first broadcasting stations, using alphanumeric codes to represent their parent papers. Over time, radio stations began employing various formats, broadcasting news, music, voice-acted theater, and even educational lectures at different times during the day. As more and more broadcasters jostled to fill the airwaves, the American government realized an agency dedicated to radio communications was needed. The Federal Radio Commission was established in 1927. It was replaced in 1934 by the Federal Communications Commission (FCC).[5]

Radio broadcasting blossomed in the prosperous economy following World War II and has remained popular ever since. Moreover, the economic boom that came after the war saw the introduction of television (TV) broadcasting to the American airwaves. Although TV had existed for decades at that point, the technology was poor and receivers were expensive.

# THE FIRST TELEVISION STATION

Have you ever wondered how old TV broadcasting is? It's older than most people think. The first TV broadcasting station was established in 1928. WRGB, based in Schenectady, New York, sent the first broadcast television signals in 1928, and it has been on the air ever since.

Earlier television signals relied on telephone wires, or even wires connecting the originator and the receiver. Radio wave broadcasting brought entertainment to households across the globe.

WRGB is the oldest television broadcasting station in the world. Although the quality of the radio signal has improved, the technology behind broadcasting is unchanged.

Technology advancements improved the picture quality and allowed the price to drop to levels most families could afford. Soon, nearly every home had a TV set along with a radio, and families gathered around their TVs to watch their favorite shows.[6]

Radio and TV broadcasting remained largely unchanged until the 1990s, when digital broadcasting was introduced. The digital signal produced a clear sound (and image, in the case of TV signals), free from static or noise. Digital transmissions began to slowly replace analog ones, until in 2009 the FCC mandated that all television signals be broadcast digitally.[7] Analog AM/FM radio signals are still broadcast, but many are broadcast digitally alongside their analog counterparts. In the future, it is likely all signals will be digital.

## Radio Wave Bands

If you have ever tuned a radio or TV to a particular station, you know that each frequency is unique. But just what are the different frequencies, and what are they for?

Radio frequencies can be grouped by their levels. This grouping is set by the International Telecommunication Union (ITU), a branch of the United Nations that oversees radio communication. Each set of radio frequencies is called a band. The table of radio bands on the next page shows their frequency and wavelength ranges.

| ITU Band | Frequency and Wavelength Range |
|---|---|
| Extremely low frequency (ELF) | 3–30 Hz<br>100,000–10,000 km |
| Super low frequency (SLF) | 30–300 Hz<br>10,000–1,000 km |
| Ultra low frequency (ULF) | 300–3,000 Hz<br>1,000–100 km |
| Very low frequency (VLF) | 3,000–30,000 Hz<br>100–10 km |
| Low frequency (LF) | 30–300 kHz<br>1,000–100 m |
| Medium frequency (MF) | 300–3,000 kHz<br>1,000–100 m |
| High frequency (HF) | 3–30 MHz<br>100–10 m |
| Very high frequency (VHF) | 30–300 MHz<br>10–1 m |
| Ultra high frequency (UHF) | 300–3,000 MHz<br>1–0.1 m |
| Super high frequency (SHF) | 3–30 GHz<br>100–10 mm |
| Extremely high frequency (EHF) | 30–300 GHz<br>10–1 mm |
| Tremendously high frequency (THF) | 300–3,000 GHz<br>1–0.1 mm |

Low-frequency bands have very limited uses, as they cannot transmit human speech. ELF, SLF, and ULF bands are used for submarine communication. VLF is used for time and geophysics signals, such as the signal that keeps cell phone clocks correct or the signal in a GPS. LF is used for ham radio, as well as navigation and time signals. MF is used for AM radio, as well as ham radio. HF is used for CB radio, radio frequency identification (RFID) signals, and radar. VHF is used for FM radio, ham radio, TV, air traffic control, land mobile radio (such as walkie-talkies), and weather radio). UHF is used for TV, mobile phones, Wi-Fi, Bluetooth, GPS, remote control systems, and satellite radio. SHF is used for Wi-Fi, radar, and satellite TV. EHF is used for Wi-Fi and military crowd control weapons. THF is used in very specialized scientific fields.

You may have noticed that some uses, such as TV broadcasting, span more than one band. These applications are in high demand and have many uses, and because their signals are not harmed by spanning multiple bands, this is allowed by the FCC.

Importantly, because many people use devices that transmit within the same bands, sometimes interference can occur. This is less common with digital transmission than with analog signals, but devices such as cell phones, remote control systems, and strong AM/FM broadcasts can interfere with devices such as walkie-talkies, causing static or even completely overriding the signal.

## Digital Broadcasts

If you were born in the 1990s or later, you may not remember a time before digital radio and television. Yet for decades, analog signals were the only signals. They were vulnerable to noise, static, and interference. People used large, complicated antennas, often covered in aluminum foil, to capture poor signals and decrease noise.

The digitization of radio broadcasts uses the same principles as amplitude and frequency modulation. However, instead of sending a carrier wave across the entire range of the wave, the digital signal is modulated to a simple binary system of on/off.

In order for a digital signal to be interpreted, a digital receiver must be used. This receiver can absorb the radio waves and translate the digital signal into a form that makes sense. Before all broadcasts were required to be digital, many types of metals could be used as antennas, and a simple connection to a speaker was all that was required to listen to the broadcast, although a television set was still required in order to see the picture. Some people could even pick up broadcasts very faintly through the metal fillings in their teeth! A digital signal, however, comes through only if a digital receiver is available to decode the message.

Modern digital audio broadcasts are now referred to as HD radio. This is a trademarked term that refers to the digital version of a broadcast. Unlike high-definition (HD) video, in which the

Many radio receivers are now digital receivers, which not only decreases static, but also allows receivers to display information such as the station, song title, and artist name.

picture is very highly detailed, HD radio does not represent an increase in detail compared with analog radio. However, because signals broadcast under the name HD radio are digital signals, they suffer from less noise and interference than analog signals. This means that the sound is cleaner, with less static or loss of fidelity. Unlike television broadcasts, there is no government mandate that audio broadcasts should be digital.

CHAPTER FIVE

# SPACE AND RADIO WAVES

People on Earth have been using radio waves to communicate for just over one hundred years. But radio waves have been all around us since the dawn of time. In fact, radio waves are naturally emitted by many objects and propagate through space constantly. It was only once humans developed technology that could absorb and record these waves that their existence opened new frontiers in research.

## What Emits Radio Waves?

In 1931, Karl Jansky was working at Bell Telephone Laboratories in New Jersey. He was responsible for eliminating static in telephone signals. He was frustrated by a repetitive static signal that didn't seem to have an obvious source. This static, he found, was coming not from human-made devices, but instead came from space. In fact, he pinpointed the source of the interference—the constellation Sagittarius.[1]

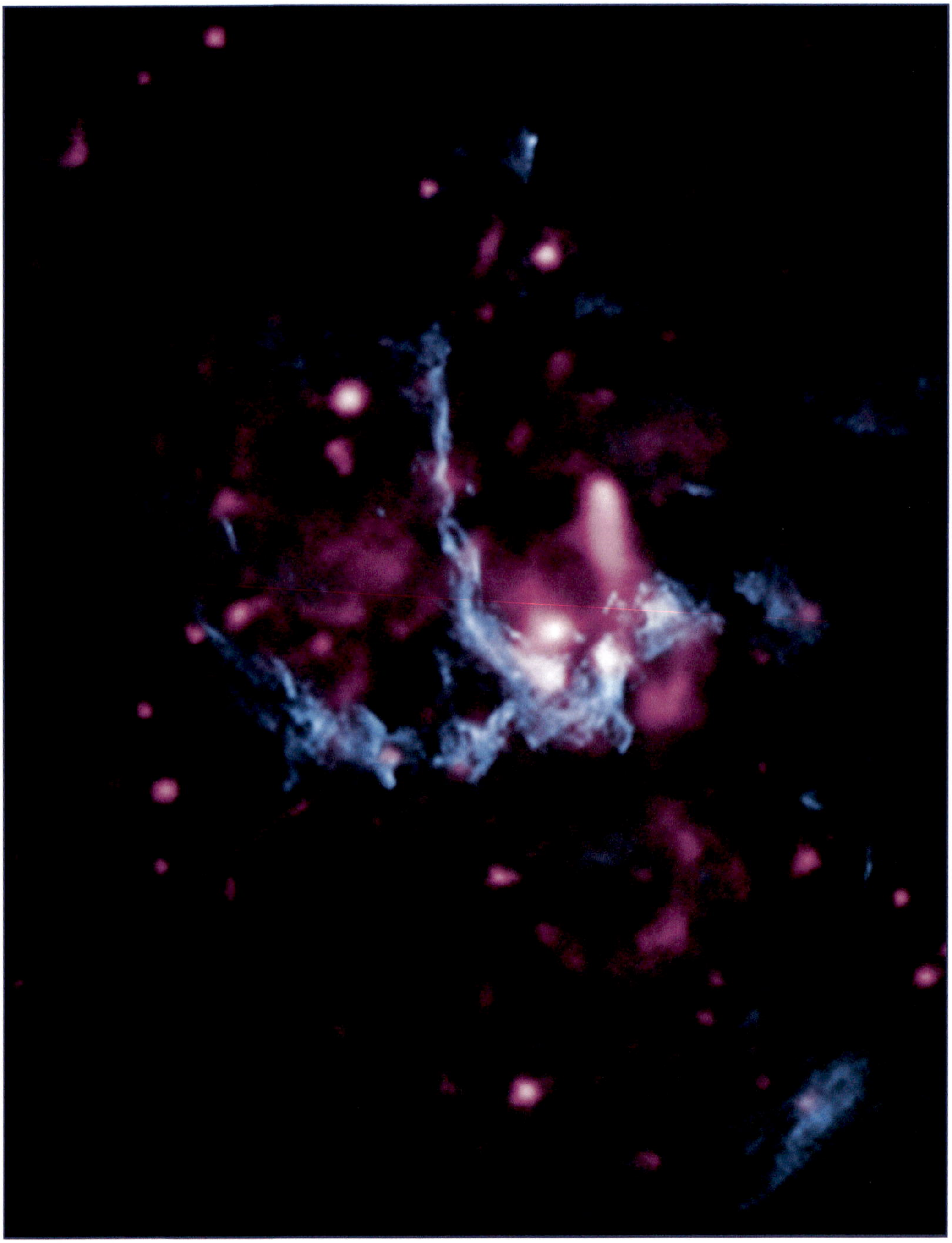

This image of the center of the Milky Way galaxy is a visual interpretation of signals received by telescopes. Radio waves are shown in blue, while X-rays are shown in purple.

Jansky had discovered radio astronomy. This field of astronomy is based on the fact that astronomical objects emit radio waves. Scientists have discovered many cosmic objects that emit radio waves, some of which they did not even know existed prior to the development of radio astronomy.

Our sun is the closest object that emits radio waves. These are the strongest radio waves that can be detected in the sky simply because the sun is the closest object emitting radio waves. Detection of the variations in these waves allows scientists to observe sunspots—places of increased solar radiation on the sun's surface. Bursts of radio waves have also been detected in conjunction with solar flares. In fact, solar flares can emit so much radiation in the radio spectrum they can disrupt radio communications, such as satellite and broadcasting signals, for hours. They are so powerful they can disrupt electrical plants![2]

Sagittarius, observed to emit radio waves by Jansky, is actually the center of the Milky Way galaxy. The source of the radio wave emission from Sagittarius appears to be a supermassive black hole, which constantly emits radio waves.[3] Radio waves have also been observed in the remnants of supernovas—stars that have exploded violently.

Pulsars and quasars were both discovered using radio astronomy. A pulsar is an extremely dense, rotating star that emits a beam of radiation. The beam can only be detected when it is pointed at the receiver, resulting in the stars having a pulsating appearance—hence the name pulsar. A quasar is a bright center of a galaxy, thought to likely be a supermassive black hole. They appear bright because their tremendous gravity pulls in all nearby matter, including burning gas and small particles. Quasars emit radiation as they pull gas into their centers.

Places where stars are coming into existence also emit radio waves. In these places, gas molecules swirl and collide with a great deal of energy, emitting light across the electromagnetic spectrum. Dense clouds of gas within galaxies, such as along the arms of the Milky Way, emit similar spectra.

## Radio Telescopes

Once radio waves had been observed emitting from cosmic sources, people began building telescopes to capture this radiation. Such telescopes are called radio telescopes. Radio telescopes can be used to observe radiation that cannot be detected by other types of telescopes.

Radio telescopes take many forms. Some look like arrays

### SUPERMASSIVE BLACK HOLES

**Supermassive black holes are thought to be the centers of all large galaxies. This is because they are the only things with enough mass and density to have sufficient gravitational pull to bring together the billions of stars that reside in galaxies.**

**Supermassive black holes get their name from their size. They are billions of times as massive as our sun. They are called black holes because nothing can escape their gravitational pull, not even light. But they are not holes. They are, in fact, the most massive stars in the universe.**

of antennas: metal rods strung together in geometric patterns. Others look like big curved dishes.

Most radio telescopes can be rotated so they can focus on a signal. The dish telescopes use their curved shape to focus the radio signal. The largest dish radio telescope is located in China. It is 0.3 miles (500 m) across![4] There are also groups of radio telescopes that can be used together to capture signals. The Very Large Array (VLA) in New Mexico consists of 28 dishes and spans over 20 miles (32 km).[5]

There are also radio telescopes in space. Earth's ionosphere,

Radio telescopes on Earth are shaped like large dishes. The dish shape helps collect radio waves and focus them to a point in the center, where the antenna can absorb them.

which is the outer layer of Earth's atmosphere that consists of ions, reflects many radio waves, especially low-frequency waves. In order to detect these waves, satellites with radio antennas have been launched into orbit around Earth. There have been three orbiting radio telescopes, although two have ended their operations, leaving only one today. However, there is a radio telescope on the side of the moon that faces away from Earth.

## Interpreting Radio Signals from Space

Radio signals from stars and cosmic gas don't reach receivers and play a tune. So what are they like, and how do scientists interpret them?

Most radio telescopes send the signals they receive to computers. These computers translate radio signals into visual representations. Radio waves are light, so the representations really do show how "bright" the object is—in radio waves rather than visible light waves.

Because radio signals can be very faint and hard to focus on, radio astronomy often uses interferometry. This technique combines signals from two or more telescopes at distant positions, which are connected by cables. The pattern made by focusing on a single spot from two different vantage points shows interference, or places where the radio waves overlap. This allows scientists to more precisely pinpoint exactly where an object is in space, as well as understand its brightness.

Not all radio waves are interpreted as pictures, though. Some

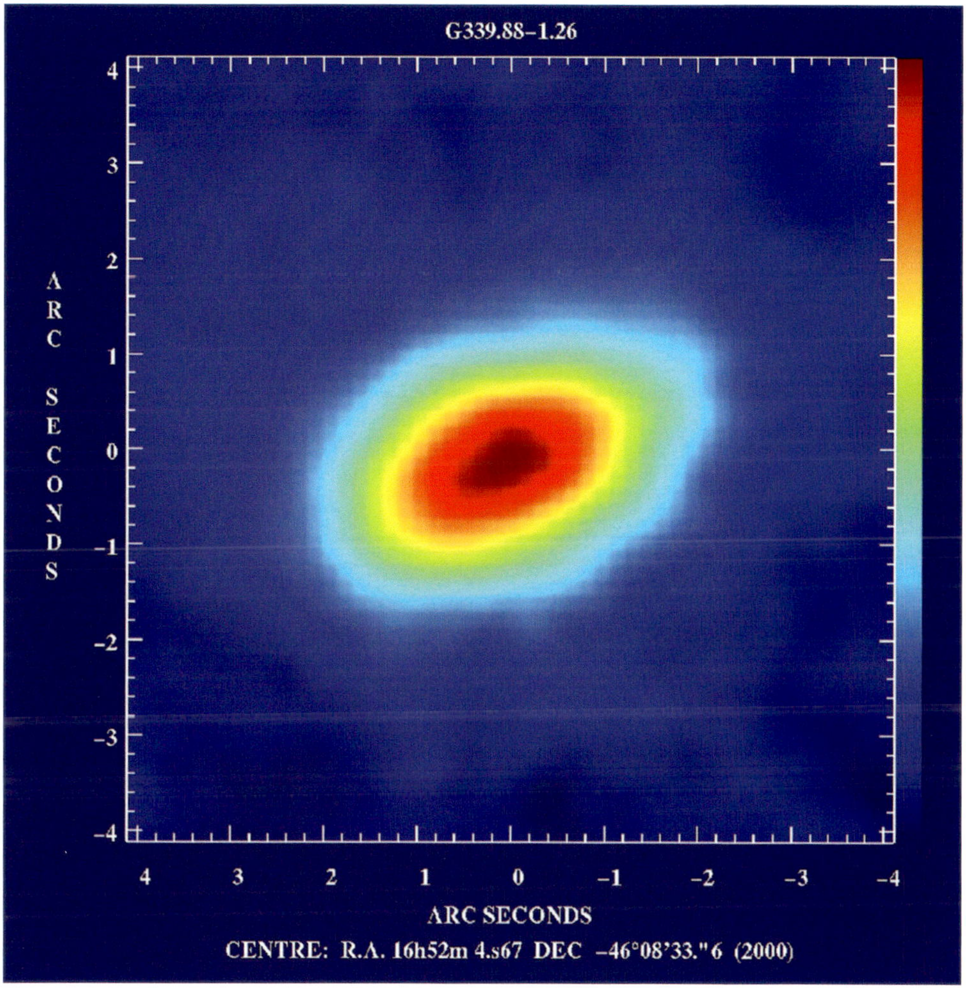


This image shows radio waves given off by an enormous star 10,000 light years away from Earth. The star is likely 20,000 times larger than the diameter of Earth's orbit around the sun!

computers instead render them as sound. These sounds can help scientists understand the size of a star and how fast it is spinning. The sounds can seem very much like music because they often have a regular rhythm based on the spin of the star.

CHAPTER SIX

# HUMANITY'S SEARCH OF THE SKY

sky and wondered, "Are we alone in the universe?" Over the past decades, astronauts have walked on the moon and scientists have landed robots on Mars and crashed probes into Jupiter. Scientists have peeked into every distant corner of our solar system, but they have not found even a speck of life. So are we truly alone?

## Sending Signals into Space

Radio waves, like all other forms of light, travel from their source through space until they are absorbed by an object. This means every radio signal ever sent into space is either still traveling or has hit some cosmic object. Every satellite broadcast, every TV show and song that left the atmosphere will continue forever until it is intercepted.

Knowing this, many people began deliberately sending messages into space, hoping to contact extraterrestrial intelligent life. Even the National Aeronautics and Space Administration (NASA) got in on the action. The Voyager

The Voyager spacecraft, both launched in 1977, are on interstellar missions. Voyager 1 left our solar system on August 25, 2012. Voyager 2 followed on November 5, 2018.

spacecraft, probes designed to leave our solar system, are each equipped with radio transmitters, constantly sending signals out in every direction.

Even without intending to send signals into space, humanity continues to do so. Communications between NASA and astronauts on the International Space Station, as well as those from similar organizations in other countries, speed out into the

## CHATTERING EARTH

**A common misconception holds that every radio transmission ever sent by anyone on Earth is still zooming through space. In fact, most radio transmissions don't reach very far. This is because materials on Earth can absorb them.**

**Furthermore, Earth's atmosphere absorbs and reflects radio signals. The ionosphere is made up of the outer two layers of Earth's atmosphere: the exosphere and thermosphere. This part of the atmosphere reflects or absorbs most electromagnetic radiation, including radio waves. So most of radio transmissions don't extend beyond the ionosphere, and most radio waves that move through space never reach Earth's surface.**

cosmos.

Ever since we learned how to send radio signals, Earth has lit up the radio sky more brightly than ever.

## Life on Other Planets

In all our searching, we haven't yet found even microscopic life anywhere but Earth, much less intelligent life. And would we recognize it even if we did find it?

The search for extraterrestrial intelligence (SETI) is a distributed network of computers that analyze the tremendous amount of radio signals received by radio telescopes, looking for evidence of intelligent communication. SETI relies on a neural network formed by linking thousands of computers. The combined computer power can scan a great deal more data than any single human could.

SETI also looks for potential extraterrestrial homes for humans. Planets are far too small to see with telescopes, but as planets pass in front of the stars they orbit, they block some of the light emitted by those stars. In this way, we have discovered thousands of planets, called exoplanets, orbiting other stars.

The most remarkable find in the search for extraterrestrial intelligence was called the Wow signal, received in 1977 at an Ohio observatory. This was a high-intensity signal transmitted from an area in the Sagittarius constellation. It was so unusually

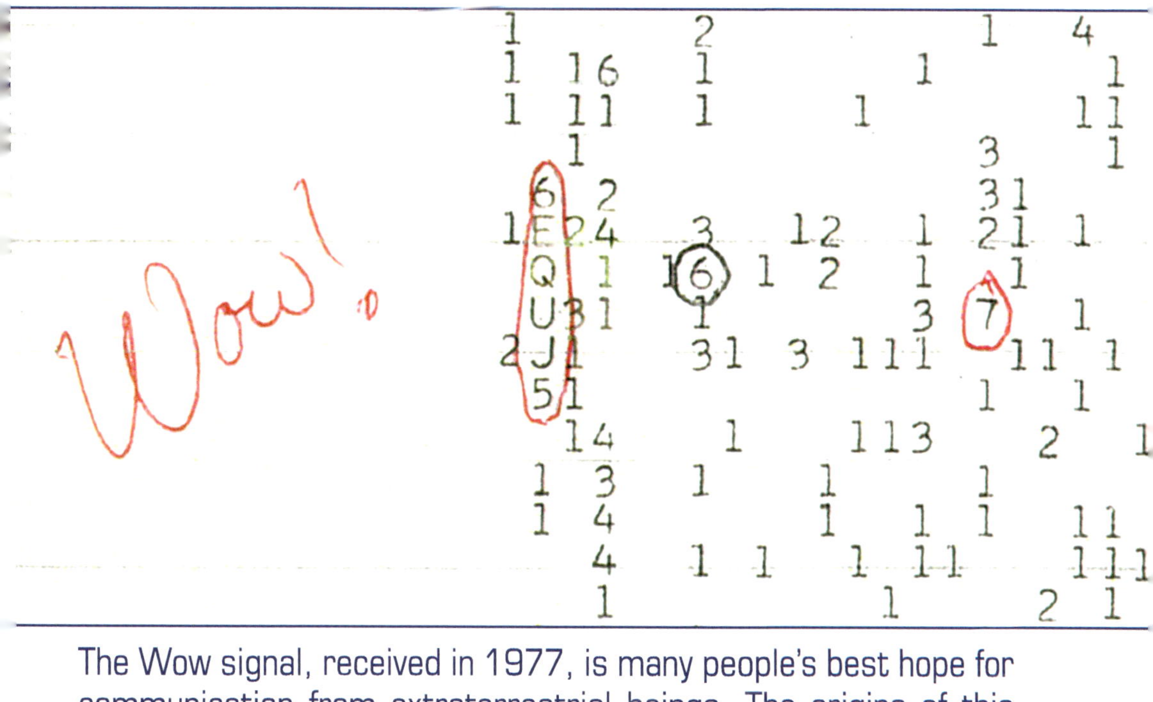


The Wow signal, received in 1977, is many people's best hope for communication from extraterrestrial beings. The origins of this anomalous signal are still unknown.

intense that the researcher who first saw it wrote "Wow!" in the margin of the paper—hence the name Wow signal.[1] Many ideas were put forth about what caused the signal, including a spinning star and a human signal bouncing off the atmosphere. A previously unknown comet seems to be one of the most likely explanations. Nothing like the Wow signal has ever been received again, despite telescopes all over the world repeatedly looking for it.

## The Fermi Paradox

The great question, then, is this: If there is intelligent life elsewhere in the universe, why haven't we heard from them? Humans are constantly sending radio communications out into space, yet they do not receive an answer in return. This silence is despite mathematical probability that thousands or even millions of Earth-like planets exist. This is known as the Fermi paradox, named after Enrico Fermi, the physicist who first described it.

Fermi's argument claims that since billions of stars like our own sun exist, and Earth-like planets orbit many of them, life should have evolved somewhere else in the universe in much the same way it evolved here. Thus, intelligent beings should be sending out radio waves the same way people on Earth do, or possibly even crossing galaxies to meet us. They should, at the very least, be hearing our own radio communications and responding to them.[2]

Yet, with the possible exception of the Wow signal, no one

has responded to our questions. Why not?

Many solutions to Fermi's paradox have been suggested. Perhaps, for some reason, our estimates are wrong. Despite there being planets in the right places orbiting the right kinds of stars, maybe the sparks needed for life to evolve just the way it did on Earth are truly very rare or even unique. Maybe there really isn't anybody else out there.

Or maybe there is, or will be, but we were first. People on Earth discovered radio waves only a mere century and a half ago. Why should other civilizations get there before us?

Perhaps another intelligent species does exist, but they are just too far away to hear. Although radio signals travel until they hit something, space is, indeed, quite full of things: stars, clouds of gas, comets, and black holes.

Possibly, suggest some, there are many, many intelligent species on multiple planets, and they all consider Earth and its creatures a protected environment. After all, we do this ourselves to species here on Earth. We set aside protected land where humans cannot hunt or study animals in order to maintain the wildness of their ecosystems. Maybe aliens are invested in protecting us from the harm they might do.

Some people take a darker view. Perhaps any species intelligent enough to achieve space travel also develop the means to destroy themselves. After all, humans went to space only a few years after making atomic bombs. Humanity stands perilously at the edge of nuclear annihilation, humanity's fate in the hands of only a few world leaders. Technology, it may be reasoned, destroys us. The evidence of global climate change emboldens this view.

Whatever the reason, one thing remains clear: As far as we know, we are alone. But that doesn't mean we won't stop searching the sky.

# OTHER USES OF RADIO WAVES

CHAPTER SEVEN

You know radio waves are used in communication. They are used to study space and the objects in it. But there are even more ways to use this form of light.

## Radar

Did you know that the word "radar" is an acronym? It is a word formed from the first letters of words that describe it. "Radar" stands for RAdio Detection And Ranging. Radar uses radio waves. It's right there in the name!

Radio relies on the fact that radio waves, like other kinds of light, reflect off of surfaces. A radar transmitter sends radio waves of a specific frequency out from a certain point. These waves reflect off of objects in front of the transmitter and travel back to it. A receiver absorbs these waves. The distance to objects can be calculated by timing the difference between the waves

The domed devices on this ship are radar transceivers. Radar helps ships navigate waters where hazards may be difficult to see with the naked eye.

being sent out and the waves being received. This calculation is usually done by a small computer in the radar device itself.

Radar has many uses. A lot of people are familiar with police radars. These devices measure the speed of oncoming cars. They can be used to let police officers know whether or not a driver is obeying the speed limit.

Nearly all boats and ships are also equipped with radar devices. These devices help ships avoid crashing into each other

## RADAR IS WAR TECHNOLOGY

**Radar is itself a technology born of war. Specifically, radar came into widespread use during World War II. Using the foundations laid by Heinrich Hertz, nearly every country involved in the war, on both sides, secretly developed and deployed radar devices.**

**Every country using radar thought it would gain a tactical advantage. Yet with all of them using it, nobody came out ahead. It was only after the war that radar's many other uses, such as ship and aircraft navigation, became widespread as well.**

and help people steering ships near shore avoid objects such as rocks and piers. Because ships often travel in the dark or in foggy conditions, radar is vital to helping them navigate without crashing into something.

Radar is also used in air traffic control to keep track of the movements of aircraft near an airport. This technology helps air traffic controllers guide aircraft as they take off and land.

The military uses radar as well. It can spot aircraft, ships, and cars with radar devices.

However, using radar to find objects is not foolproof. Objects can be designed to "hide" from radar. This is especially common in vehicles, notably military aircraft. By reducing sharp angles and building the craft out of radio-absorbent materials, these stealth planes can be made to be difficult to detect even with sophisticated radar devices.[1]

## Submarine Communication

The military has many applications for radio: short-distance communication between small groups, long-range broadcasts, and radar, just to name a few. Military watercraft need to be able to communicate with each other and with commanders on land. But salt water, which makes up the oceans, stops most radio waves from propagating. And submarines can remain submerged for months at a time. But submarines, like surface ships, need to communicate.

One solution to help submarines solve this problem is simply to surface every now and then to send and receive radio signals. However, this may mean that they miss important transmissions that occur when they are submerged. Additionally, submarines are most vulnerable at the surface, where they can be spotted and attacked by both surface ships and aircraft. This solution is not ideal.

Instead, submarines rely on VLF and ELF radio waves for communication. These waves, with their low frequencies, are able to propagate through seawater when higher-frequency waves cannot. VLF waves can penetrate seawater several meters below the surface, while ELF waves can reach even deeper—hundreds of meters deep.[2]

This communication is not as simple as using a walkie-talkie, however. For one thing, the antennas required for broadcasting these waves must be quite large, so the broadcasts must be one-way only. They are sent from shore to ship, with no ability for the submarine to respond. Furthermore, building these antennas is quite an endeavor. The wavelength of waves in the ELF range is half the diameter of Earth. This is no small feat for construction! To tackle this problem, transmitters employ multiple stations separated by large distances, connected by wires. However, building antennas in this manner is costly and risky from a tactical standpoint.

Compounding the problem, human voice cannot be transmitted on these frequencies. Thus all information sent via these waves is encoded as symbols. In many ways, this offers a benefit. Such encoding means that intercepted transmissions must be decoded before they are interpreted, and military enemies do not have access to each other's coding cyphers.

## Wi-Fi and Mobile Phones

Radio communication commonly takes a form most people likely do not think of when they think of radio waves. All wireless fidelity (Wi-Fi) signals use radio waves.

The bands allocated for Wi-Fi signal are referred to as the 2.4 GHz and 5 GHz bands. These include all frequencies between 2,400–2,500 MHz and 5,725–5,825 MHz, respectively. Originally, only the 2.4 GHz band was used, but as more and more Wi-Fi devices were made, the 5 GHz band increased in popularity. This was simply because the frequency was less crowded, leading to faster communication speeds and less noise. Many modern devices are called "dual band" capable, meaning they scan for the band with less traffic and choose it. A few devices use frequencies outside of these bands.[3]

Wi-Fi, as people who have ever strayed too far from a router while using a laptop know, has very limited range. At best, only about 100 feet (30 m) from the signal source is covered.[4] Many items can block the signal, including walls and large furniture.

Wi-Fi works by sending data between a router and connected devices. The router is typically connected to a wired cable modem, although in some cases, such as satellite internet, the modem itself may use radio waves to connect to the internet. The router and all devices connected to it use transceivers, which are antennas that function as both transmitters and receivers. The devices must "recognize" the router, and many router owners choose to protect their own router's signal with a password.

For mobile phones, Wi-Fi tethering is possible. Tethering occurs when one phone functions as a router, using the phone's wireless data and connecting other devices to the signal. This can be useful when access to a cable or satellite modem is not possible, although data sent using this signal tends to be significantly slower than data sent using cable modems.

Mobile phones themselves also use radio waves to work. The phone is a transceiver, sending signals to a cell tower and receiving them in return. The tower receives the signal from the phone and sends it along cables underground. In the case of data, the signal is sent to a computer server connected to the internet. In the case of phone calls between mobile phones, the signal is sent to a tower near the other phone, which again uses radio waves to send the signal to the recipient's phone. The towers also regulate which frequency each phone uses to ensure there aren't too many phones using the same frequencies and interfering with each other.

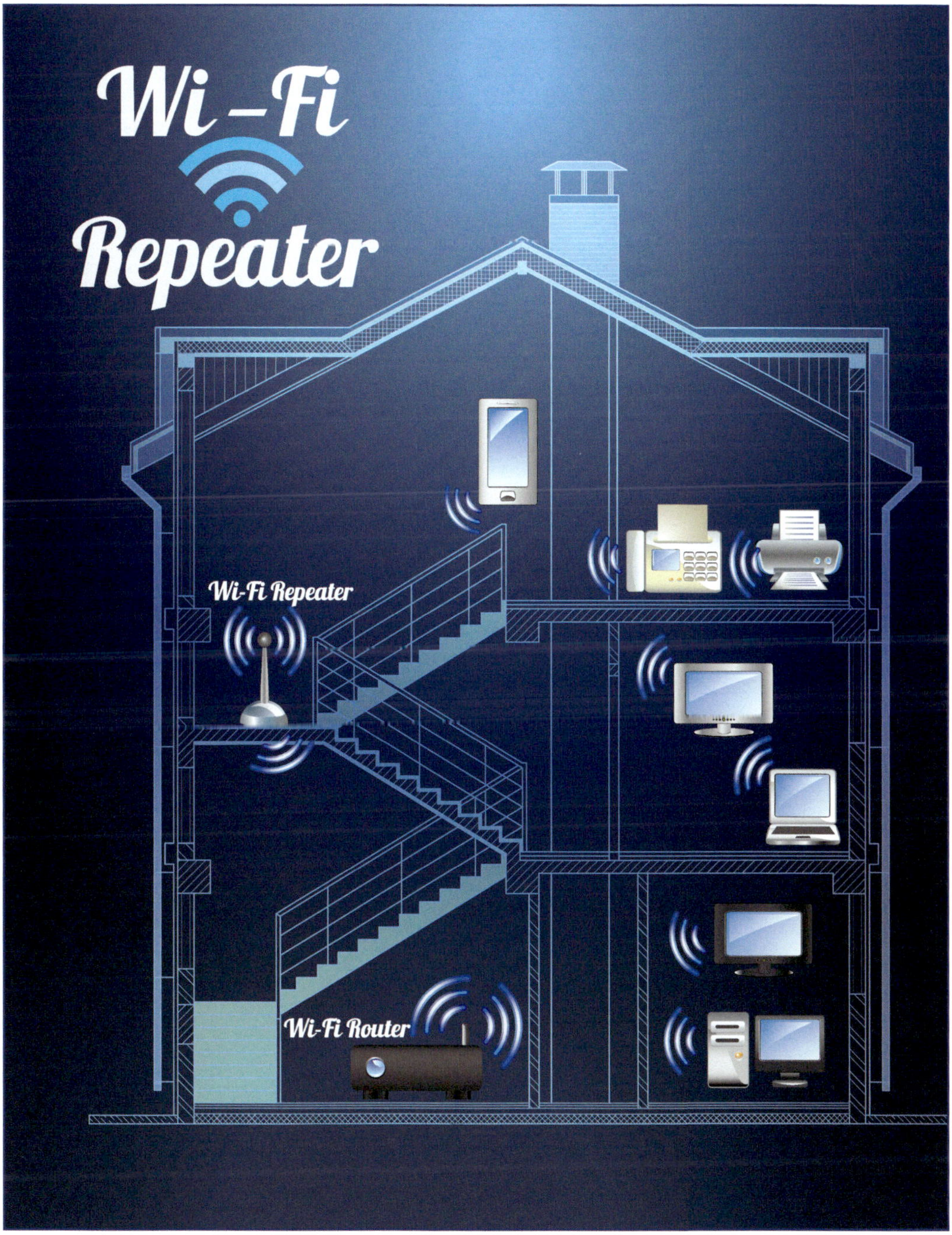


Routers rely on radio waves to communicate with devices. The farther from the router, the weaker the signal, especially if walls are in the way. Repeaters can amplify the signal for distant devices.

Since their discovery, radio waves have been one of the primary ways humans communicate with each other across vast distances. The discovery that radio waves also emanated from cosmic objects opened the skies to new knowledge and understanding. Humans, ever curious, search the cosmos for other beings who may be using radio waves the same way we are. Meanwhile, people will keep using them here on Earth to keep in touch with each other.

# CHAPTER NOTES


## CHAPTER 1
## THE ELECTROMAGNETIC SPECTRUM

1. Carlos A. Altget, "The World's Largest 'Radio' Station," High Energy Physics at UW Madison. Retrieved February 18, 2019, https://www.hep.wisc.edu/~prepost/ELF.pdf.
2. "Radio Waves," Boundless Physics. Retrieved February 19, 2019, https://courses.lumenlearning.com/boundless-physics/chapter/the-electromagnetic-spectrum/.
3. "A Brief History of Lightning Detection," NOAA SciJinks. Retrieved March 5, 2019, https://scijinks.gov/lightning-detection.

## CHAPTER 2
## THE DISCOVERY OF RADIO WAVES

1. "James Clerk Maxwell," *Encyclopedia Britannica*, online ed., s.v.
2. "James Clerk Maxwell," *Encyclopedia Britannica*.
3. James Clerk Maxwell, *Treatise on Electricity and Magnetism* (Oxford, UK: MacMillan, 1873).
4. "Herschel Discovers Infrared Light," Cool Cosmos. Retrieved February 20, 2019, http://coolcosmos.ipac.caltech.edu/cosmic_classroom/classroom_activities/herschel_bio.html.
5. "Hertz's Experiments," Do Physics Online. Retrieved February 20, 2019, http://www.physics.usyd.edu.au/teach_res/hsp/u7/t7_hertz.pdf.

## CHAPTER 3
## TRANSMITTING AND RECEIVING RADIO WAVES

1. "How Does Modulation Work?" Radio Academy. Retrieved February 21, 2019, https://www.taitradioacademy.com/topic/how-does-modulation-work-1-1/.

## CHAPTER 4
## RADIO WAVES IN COMMUNICATION

1. "Marconi Sends First Atlantic Wireless Transmission," History.com. Retrieved February 26, 2019, https://www.history.com/this-day-in-history/marconi-sends-first-atlantic-wireless-transmission.
2. "Alexander Graham Bell," History.com. Retrieved February 26, 2019, https://www.history.com/topics/inventions/alexander-graham-bell.
3. Marconi Sends First Atlantic Wireless Transmission," History.com.
4. "A Short History of Radio," Federal Communications Commission. Retrieved February 26, 2019, https://transition.fcc.gov/omd/history/radio/documents/short_history.pdf.
5. "The Radio Act," *Central Law Journal*, March 4, 1927, p. 158.
6. "History of the Television," Business Ed. Retrieved February 26, 2019, https://bebusinessed.com/history/history-of-the-television/.
7. Digital Television Transition and Public Safety Act of 2005.

## CHAPTER 5

## SPACE AND RADIO WAVES

1. "Karl Jansky," *Encyclopedia Britannica*, online ed., s.v.
2. "The Radio Sun," Radio2Space.com. Retrieved February 27, 2019, https://www.radio2space.com/the-radio-sun/.
3. "Most Detailed Observations of Material Orbiting Close to a Black Hole," European Southern Observatory (ESO). Retrieved February 27, 2019, https://www.eso.org/public/news/eso1835/.
4. "China Exclusive: China Starts Building World's Largest Radio Telescope," People's Daily Online, December 26, 2008. http://en.people.cn/90001/90776/90881/6562884.html
5. "Very Large Array," National Radio Astronomy Observatory. Retrieved February 27, 2019, https://public.nrao.edu/telescopes/vla/.

## CHAPTER 6

## HUMANITY'S SEARCH OF THE SKY

1. John Kraus, Ohio State Radio Observatory. "The Tantalizing WOW! Signal." Copy of letter to Carl Sagan containing an unpublished paper describing the event, January 31, 1994, http://jump2.nrao.edu/dbtw-wpd/textbase/Documents/kraus-wow-unpublished-30jan1994.pdf.
2. E. M. Jones, "'Where Is Everybody?' An Account of Fermi's Question." Los Alamos National Laboratory, March 1, 1985, https:www.osti.gov/scitech/servlets/purl/5746675.

## CHAPTER 7
## OTHER USES OF RADIO WAVES

1. Graham Brooker, "How Does Radar Work?" ABC Science. Retrieved March 01, 2019, http://www.abc.net.au/science/articles/2014/03/17/3964782.htm.
2. "Submarine Communications," GlobalSecurity.org. Retrieved March 4, 2019, https://www.globalsecurity.org/military/systems/ship/sub-comm.htm.
3. "Wi-Fi Channels, Frequencies, Bands & Bandwidths," Electronics Notes. Retrieved March 4, 2019, https://www.electronics-notes.com/articles/connectivity/wifi-ieee-802-11/channels-frequencies-bands-bandwidth.php.
4. "How Far Can a WiFi Signal Travel?" Actiontec. Retrieved March 4, 2019, https://www.actiontec.com/wifihelp/how-far-can-wifi-signal-travel-2/.

# GLOSSARY

**AMPLITUDE** The height of a wave, measured as half the distance from the crest to the trough.

**CARRIER WAVE** The radio wave that undergoes modulation, which does not itself carry information.

**CITIZENS BAND (CB) RADIO** Radio devices and their associated signals that can be operated without a license from the Federal Communications Commission.

**ELECTROMAGNETIC RADIATION** A form of energy without mass that behaves as both a wave and a particle; also called light.

**ELECTROMAGNETIC SPECTRUM** The entire range of all types of electromagnetic radiation, or light, including X-rays, ultraviolet light, visible light, infrared, and radio waves.

**FREQUENCY** The number of oscillations of a wave that pass a given point in a given amount of time.

**HERTZ** The unit that measures the frequency of radio and sound waves.

**INTERFEROMETRY** A technique that combines signals from two or more telescopes at distant positions that allows scientists to determine a cosmic object's location and brightness.

**MODULATION** The technique of encoding information in a radio wave by changing either its amplitude or frequency.

**PULSAR** An extremely dense, rotating star that emits a beam of radiation.

**QUASAR** A bright center of a galaxy, thought to likely be a supermassive black hole.

**RADIO WAVE BAND** A grouping of frequencies designated for a specific purpose by the International Telecommunication Union.

**TRANSCEIVER** A radio device that both transmits and receives radio signals, such as a walkie-talkie.

# FURTHER READING

## BOOKS

Barron, Andrew. *Amsats and Hamsats: Amateur Radio and Other Small Satellites*. Seattle, WA: Amazon Digital, 2018.

Hamen, Susan E. *Who Invented the Radio?: Tesla vs. Marconi*. Minneapolis, MN: Lerner Publishing, 2018.

Kurtz, Kevin. *Cutting-Edge Black Holes Research*. Minneapolis, MN: Lerner Publishing, 2019.

Ringstad, Arnold. *What's Inside a Radio?* Mankato, MN: Child's World, Inc., 2019.

Sirota, Lyn. *Information Waves*. Vero Beach, FL: Rourke Educational Media, 2016.

## WEBSITES

**How Radio Works**

*electronics.howstuffworks.com/radio.htm*

A description of the parts of a radio; the way radio devices encode, transmit, and receive information; and an explanation of the physics involved.

**Radio JOVE: NASA**

*radiojove.gsfc.nasa.gov*

A NASA program for students to build and use radio telescopes to observe radio emissions within our solar system and share their observations with others in the program.

**Teach Engineering Lesson: Riding the Radio Waves**

*www.teachengineering.org/lessons/view/duk_amradio_tech_less*

A standards-based lesson with hands-on activities, assessments, and references.

# INDEX

## S



## T



## V



## W